Hello, my name is Albert, and I am from the planet Earth, just like you.

From Earth we are able to study the Universe,
and we learn about space.
We see the moon, stars, planets,
and sometimes even comets.

We have learned many things by watching the sky. The sky gives us a way to tell directions, time, and even the date. It's been very helpful to us all.

AND WE ALSO EXPLORE SPACE
USING MACHINES.

This is the sun, and it's found at the center of our solar system. All the planet go around it.
The sun is the largest body in the solar system, and it produces the light and energy that earthlings need.

Hi Sol!

Nearby is the small and fast planet Mercury. It's always near the sun. Mercury is a rocky world with no moons or rings. Two robots visited and took many photographs. There are no lakes, trees, or life. It's just covered with craters.

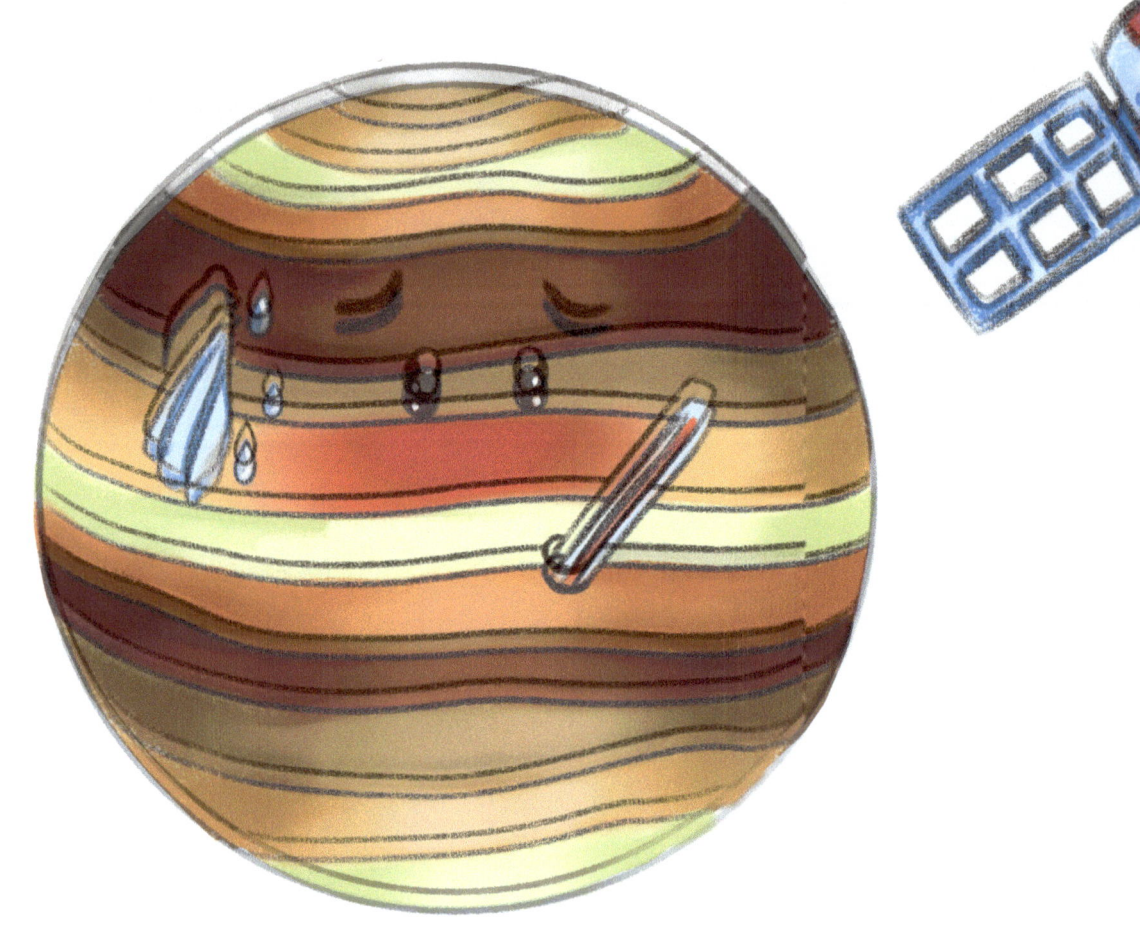

Next is Venus, and it's the hottest planet. It's covered in volcanoes. On Venus, the day is longer than the year, and it even spins backward. Many spacecraft and landers have visited Venus.

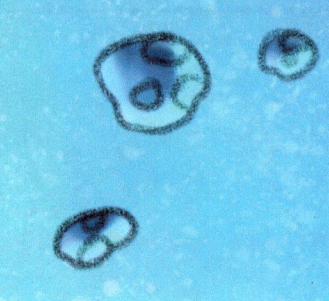

We will zip past Earth and go straight to the Red Planet Mars. It's an exciting place to visit and has had many robot visitors.
Mars even has two moons that look like potatoes. Some think there might even be
life on Mars.

Jupiter has a moon named Io. And it's the most volcanically active body in our solar system. Another moon, Europa, even has an ocean under its ice. I wonder if there are fish there too.

Here's Jupiter, the biggest planet in our solar system. It's made of gas, so there is no place for us to land. Jupiter does have many moons for us to explore, though. Some are even as big as a planet. Many spacecraft have been sent to study Jupiter.

The beautiful ringed planet Saturn. I can watch it through my telescope back home for hours. Saturn's not the only planet with rings, but I think they're the best. Saturn is the farthest planet we can easily see from Earth. We can't land on Saturn, but we can land on one of its many moons.

My favorite moon is Titan. And it's Saturn's largest. It's the most interesting of all the moons. A robot even landed there and sent back photographs.

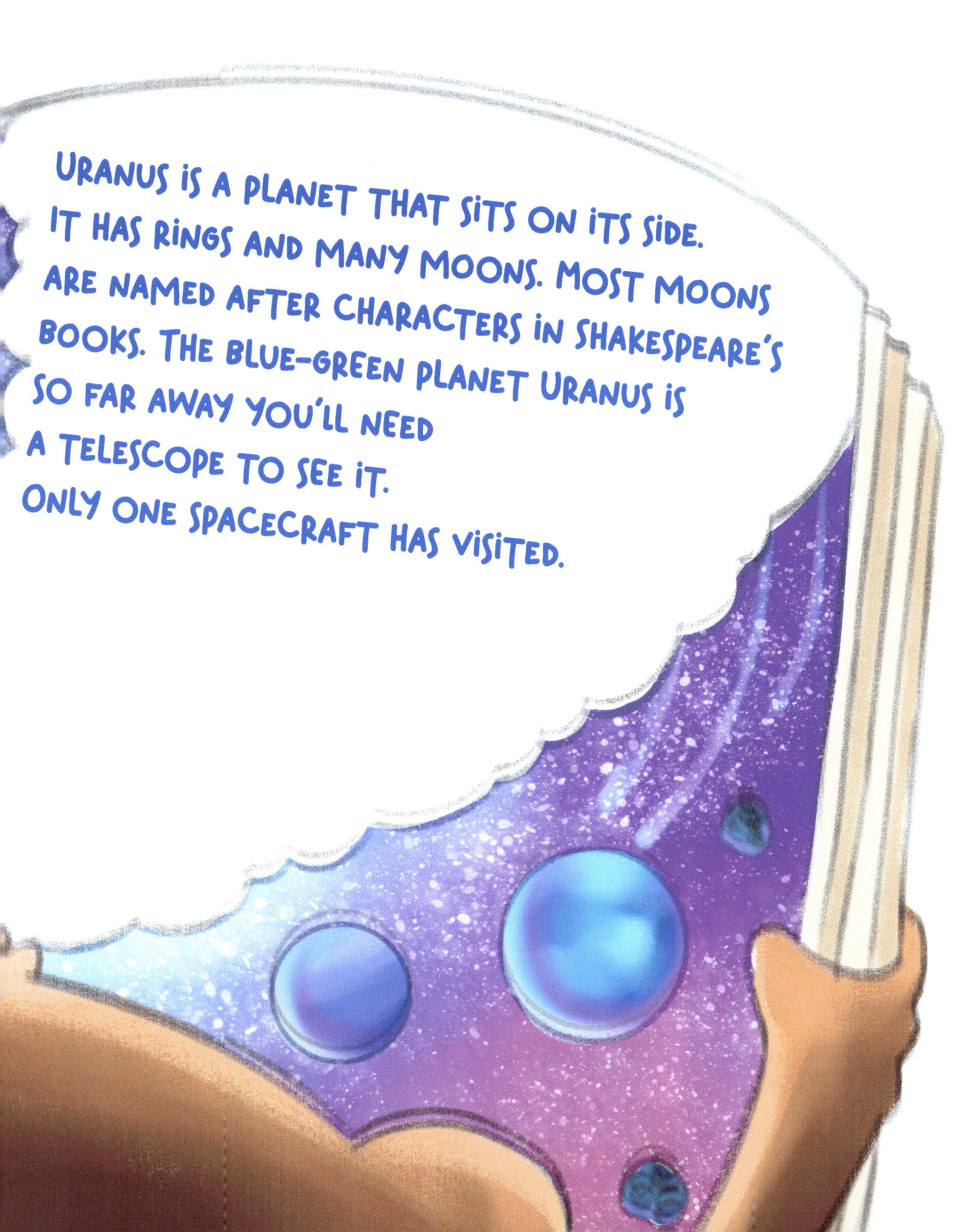

URANUS IS A PLANET THAT SITS ON ITS SIDE. IT HAS RINGS AND MANY MOONS. MOST MOONS ARE NAMED AFTER CHARACTERS IN SHAKESPEARE'S BOOKS. THE BLUE-GREEN PLANET URANUS IS SO FAR AWAY YOU'LL NEED A TELESCOPE TO SEE IT. ONLY ONE SPACECRAFT HAS VISITED.

The last of the major planets is Neptune.
It's very far away, and is a
cold, dark and windy planet.
It takes Neptune 165 years just
to go around the Sun one time.
It's an ice giant, just like Uranus,
and it also has thin, dark rings.

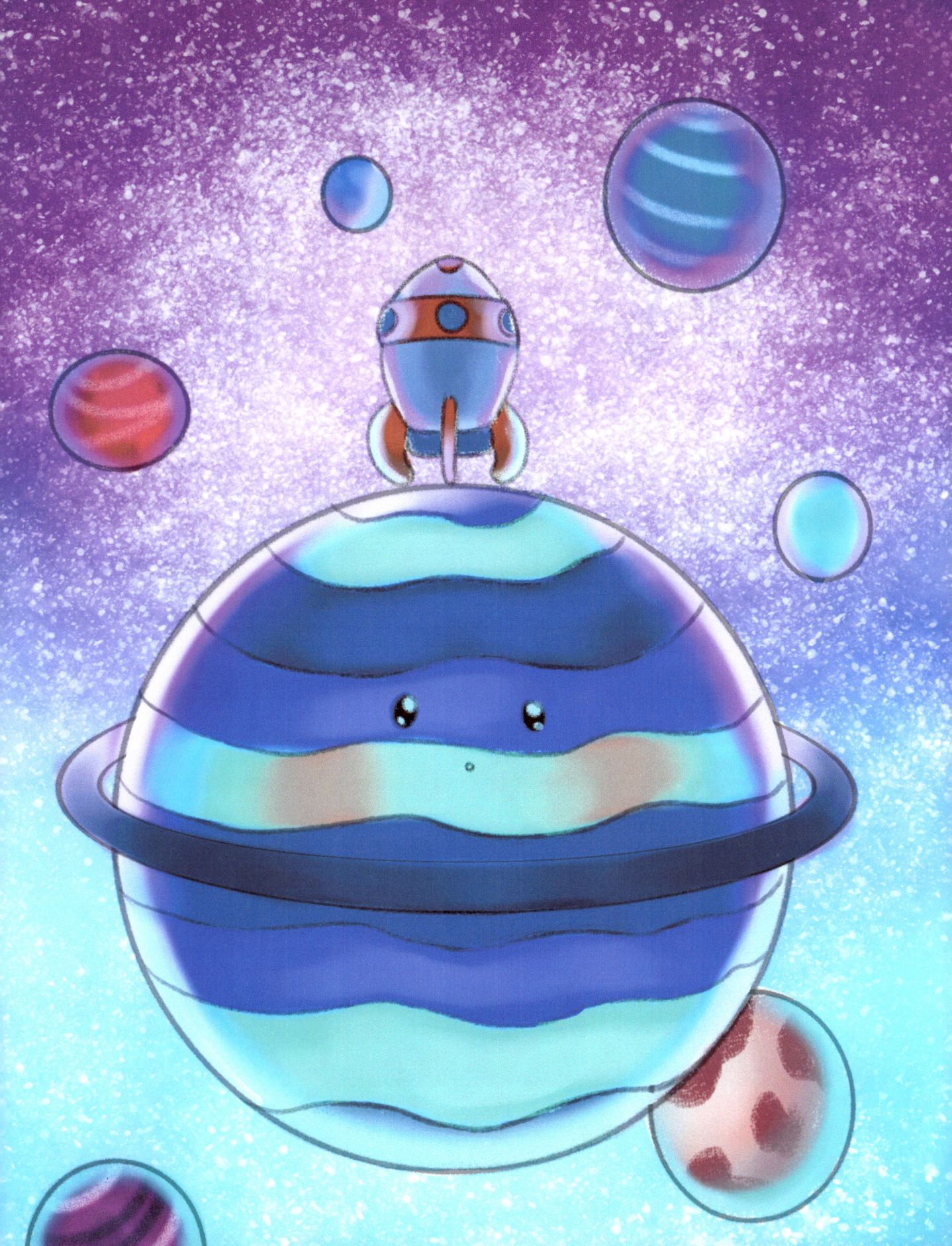

IT'S NOW TIME FOR US TO HEAD BACK HOME.

That was a great trip through the solar system, but I'm always happy to be back on Earth with you. The planet Earth is our spaceship, and it provides us with everything we need. There is no place like home, and we must take good care of it. Goodbye for now, my friend. We will meet again soon.

www.ingramcontent.com/pod-product-compliance
Lightning Source LLC
Chambersburg PA
CBHW051935210526
45473CB00006B/2260